D1370552

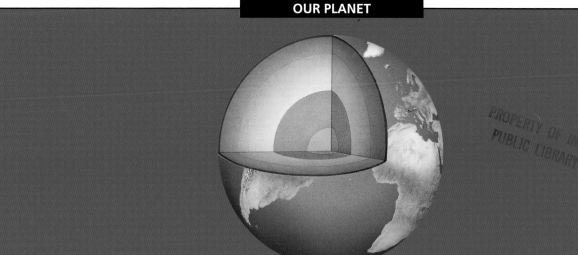

SECRETS OF THE EARTH

CHELSEA HOUSE
PUBLISHERS

A Haights Cross Communications ✝ Company ®

Philadelphia

First hardcover library edition published
in the United States of America in
2006 by Chelsea House Publishers,
a subsidiary of Haights Cross Communications.
All rights reserved.

A Haights Cross Communications ✦ Company ®

www.chelseahouse.com

Library of Congress Cataloging-in-Publication
applied for.
ISBN 0-7910-9011-6

Project and realization
Parramón, Inc.

Texts
Eduardo Banquieri

Translator
Patrick Clark

Graphic Design and Typesetting
Toni Inglés Studio

Illustrations
Marcel Socías Studio

First edition - March 2005

Printed in Spain
© Parramón Ediciones, S.A. – 2005
Ronda de Sant Pere, 5, 4ª planta
08010 Barcelona (España)
Norma Editorial Group

www.parramon.com

TABLE OF CONTENTS

OUR HOME IN THE UNIVERSE

This book is aimed at young readers who are interested in learning about the Earth. The science that deals with the Earth is called geology. Geology is an exciting field that deals with all the questions about our planet: its origin, composition, structure, present-day processes, and evolution.

This book is not meant to be a thorough reference work. Instead, we have chosen a few topics (including volcanoes, earthquakes, plate movements, and igneous and metamorphic rocks) to show that the Earth is a living planet with a dynamic interior.

We begin with an explanation of the origin of the Earth, and then discuss its structure and composition. Next, we look at its internal dynamics and their consequences (such as plate tectonics and volcanoes). Then, we explore different aspects of its external activity (action of underground waters, glaciers, and seas). Finally, we have included a chapter about topographical maps, a basic tool for the study and understanding of the Earth's surface.

It is impossible to understand present-day Earth without knowing about its past. For this reason, we look at the most important events in the evolution of our planet in the introduction, and we also include a table of geological periods at the end of the book.

A PLANET WITH A HISTORY

The Earth has a special place in the solar system. Its closeness to the sun gives it the warmth that is necessary for life to form.

SUITABLE FOR LIFE

The Earth is the third closest planet to the sun, which is about 93 million miles (150 million km) away. It is the fifth largest of the nine planets in our solar system. It has a gaseous layer, called the atmosphere, which disperses light and absorbs heat, and, in so doing, keeps it from getting too hot in the daytime or too cold at night.

Approximately 70% of the Earth's surface is covered by water, which helps regulate temperature. The water that evaporates forms clouds, which later falls back to the Earth as rain or snow, forming rivers and lakes.

As far as we know, the Earth is the only planet where life exists, where there is an oxygen atmosphere, and where there is water on the surface, although this has not always been the case.

When we discuss the history of our planet, we come across the same problem we have with the history of people: There is a long period of time about which we know nothing. This period is called the Cryptozoic Era. The Phanerozoic Period is the historical period of time, about which there is information available. This period is much shorter than the Cryptozoic Era.

THE CRYPTOZOIC ERA

This is the oldest, longest, and most mysterious part of the Earth's evolution. The word *cryptozoic* means "hidden life." The first living things appeared at the end of this period, but we have hardly any existing traces of them. This period, also called the Precambrian Eon, makes up the first 3.9 billion years of the Earth's history.

The Precambrian Eon is divided into three stages:

• **Prearchaic** (from 4.5 billion years ago to 3.8 billion years ago). It includes the first moments when the world was formed and a period when meteorites hit the Earth. At the beginning, our planet was in a molten (melted) state, and was

During the first few million years of its existence, meteorites bombarded the Earth.

The Archaic Period was a time of great volcanic activity. The gases released in these eruptions formed part of the atmosphere, and the water vapor condensed, causing heavy rains that formed the seas and oceans.

surrounded by an atmosphere that was made up of large amounts of water vapor and other gases that were harmful to living things. After tens of millions of years, the surface of the Earth and its atmosphere cooled, and much of the vapor turned to liquid, which allowed oceans to form.

• **Archaic** (from 3.8 billion years ago to 2.5 billion years ago). At this stage, the Earth's crust was hard and cool, and a primitive atmosphere already existed. There was a lot of volcanic activity, which produced huge amounts of magma that rose to the surface, cooled, and became part of the Earth's crust.

• **Proterozoic** (from 2.5 billion years ago to 530 million years ago). The first organisms able to breathe and use oxygen appeared, as did the first eukaryotes (bacteria and some photosynthetic green algae). It was during this period that the primitive atmosphere turned into an oxygen-rich atmosphere, which allowed these early life-forms to diversify and evolve into most of the invertebrates (animals without a hard skeleton).

THE PHANEROZOIC EON

The term *Phanerozoic* comes from the Greek language and means "revealed life." This period began approximately 570 million years ago with an explosion of life, which then spread and diversified throughout the oceans. During this time, the continents and oceans as we now know them formed. The evolution of living things also took place, leading to the appearance and development of human beings. The Phanerozoic Eon is divided into three stages: Paleozoic (from 570 million years ago to 225 million years ago); Mesozoic (from 225 million years ago to 65 million years ago); and Cenozoic (from 65 million years ago to the present day, including the Quaternary Period, in which human beings appeared).

In the Proterozoic Era, living creatures diversified greatly. All of these life-forms were marine invertebrates that lacked a hard skeleton. Because these creatures had no hard skeleton to leave behind when they died, there are few traces of these animals in the fossil record.

• **Paleozoic.** In this period, two important mountain-building events took place: the Caledonian and the Hercinian tectonic movements, which caused the breakups and collisions of continents and led to the formation of mountain ranges. Scientists have learned that there were two great glaciations, with temperate periods in between. The colonization of the continents, at first by plants and arthropods, and then by vertebrates (animals with backbones) such as amphibians, happened during this time.

In the Mesozoic Era, reptiles lived in all the ecosystems and came to dominate life on the planet for around 200 million years. Many of them went extinct at the end of this period, and mammals took their place.

• **Mesozoic.** During this period, the landmasses that formed the supercontinent of Pangea separated, but there was no major mountain-building event. The climate was hot and tropical worldwide for most of the period. Reptiles became widespread, and mammals and birds appeared. The first flowering plants also arose. The Mesozoic Period ended with the extinction of the dinosaurs.

• **Cenozoic.** This period, meaning "new animals" in Greek, began 65 million years ago. It includes the Tertiary Period (from 65 million years ago to 2.5 million years ago) and the Quaternary Period (from 2.5 million years ago until the present day). During this time, the continents took on their present-day shape. Another feature of this era is the spread of mammals, which replaced reptiles as the main forms

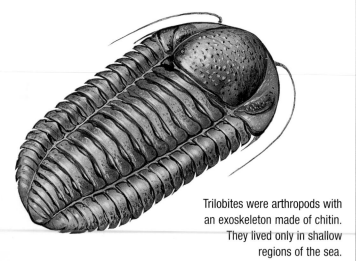

Trilobites were arthropods with an exoskeleton made of chitin. They lived only in shallow regions of the sea.

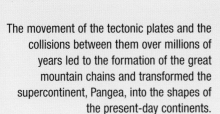

The movement of the tectonic plates and the collisions between them over millions of years led to the formation of the great mountain chains and transformed the supercontinent, Pangea, into the shapes of the present-day continents.

of life on Earth. In the middle of the Cenozoic Era, 30 million years ago, the first primates with prehensile (able to grasp) hands appeared. Various kinds of monkeys, as well as human beings, are part of this order of mammals, although hominids did not appear until the end of this era, during the Quaternary Period. During this time, the Earth experienced a number of spectacular mountain-building events.

Today, the Earth is still in the process of transforming itself. The crust is broken into 15 constantly moving enormous plates, whose edges are always being created or destroyed. Thus, within another 150 million years, Africa will have divided in two, and one of these parts will have moved north to join Europe. Antarctica will merge with Australia, and California will move north to collide with Alaska.

The first hominids (members of the primate family) appeared at the end of the Tertiary Period, but it was not until very recent times, during the final stage of the last ice age (40,000 years ago), that *Homo sapiens* (modern humans) appeared on Earth.

A SLOW SERIES OF EVENTS

The Earth formed 4.6 billion years ago from a cloud of gas and cosmic dust that made up the origin of the solar system. At the beginning, it was a half-molten ball of matter. The heavier elements fell toward the center to form the metal core; the lighter elements moved to the surface to create the rocky mantle and the crust. Over billions of years, the planet continued to cool, the surface turned solid, and the atmosphere and the oceans were formed.

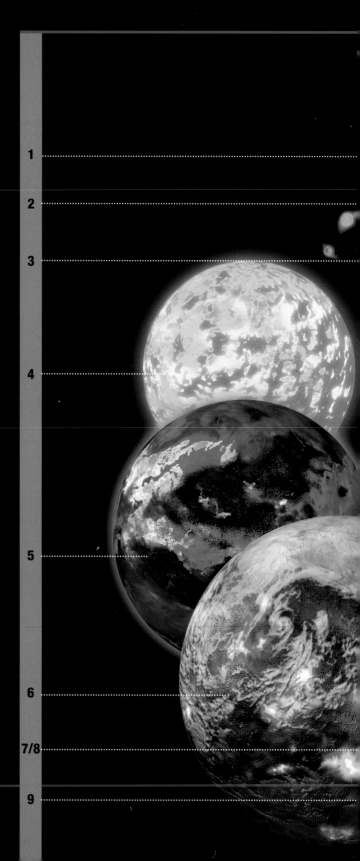

1 cloud of gas and dust

2 formation of the sun
4.6 billion years ago, a cloud of gas and dust contracted to form the sun

3 formation of the planets
other parts of the cloud of gas and dust formed solid lumps of ice and rock that joined to form the planets

4 radioactivity
radioactivity in the rocks made the newborn Earth melt at the beginning

5 core formation
iron and nickel sank to form the Earth's core, while there was an ocean of molten rock on the surface

6 great volcanic activity
the crust grew in size and the atmosphere was enriched with gases

7 formation of the atmosphere and the oceans
gases from volcanic eruptions began to form the atmosphere, and water vapor condensed to form the oceans

8 3.5 billion years ago
the greater part of the crust had already formed, but the appearance of the continents was very different from what we know today

9 the present day
the crust has broken up into huge plates whose edges are constantly being created and destroyed

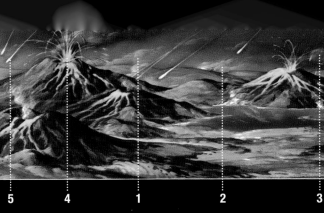

5   4   1   2   3

1 water evaporates and condenses; **2** condensation of water vapor;
3 abundant rains and lightning; **4** expulsion of magma and gases;
5 rain of meteorites

The primitive crust

As the crust solidified, water vapor began to condense and fall in the form of rain, causing the first oceans to form. Intense volcanic activity provided enough water vapor and gases to form the early atmosphere.

THE PRIMITIVE ATMOSPHERE

The Earth's primitive atmosphere was composed of methane (CH_4), ammonia (NH_3), sulphydric acid (H_2S), carbon dioxide (CO_2), and water vapor (H_2O). The hydrogen and helium escaped into space, and free oxygen appeared millions of years later, when the first photosynthetic organisms appeared.

THE MOVEMENT OF OUR PLANET

The main cause of meteorological (weather-related) and climatic variations is the Earth's orbit around the sun. Our planet has two main astronomical movements: rotational and orbital. These movements cause the seasons, the succession of days and nights, and the temperature differences on our planet.

1 orbit
in its movement around the sun, the Earth has an elliptical orbit. It takes more than 365 days to make a complete trip around the sun

2 rotational movement
is the movement of the Earth around its own imaginary axis; it takes 24 hours to make a complete spin from west to east

3 winter solstice
takes place from December 21 to December 22. The rays of the sun reach their maximum inclination with respect to the Earth's axis. The North Pole is at its farthest distance from the sun, so winter begins in the Northern Hemisphere, and summer begins in the south

THE DISTANCE TO THE SUN CHANGES

Because the orbit of the Earth around the sun is elliptical (oval-shaped), the distance between the two bodies varies according to the time of year. When we are closest to the sun (in the month of January), a distance known as the perihelion, we are 91.8 million miles (147.7 million km) from the sun. At the farthest distance (in the month of July), a distance called the aphelion, we are 94.6 million miles (152.2 million km) from the sun.

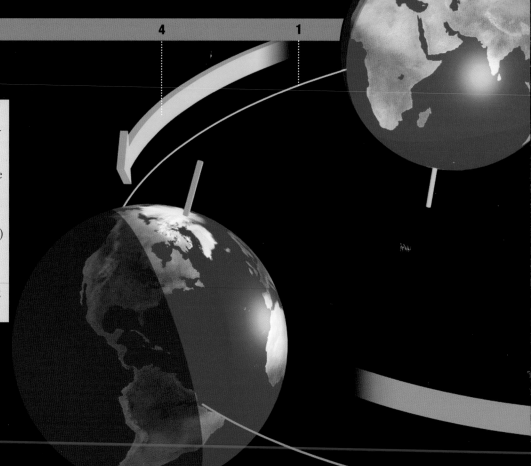

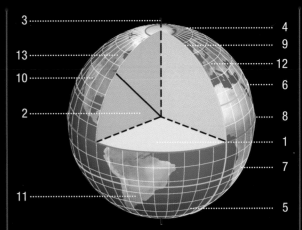

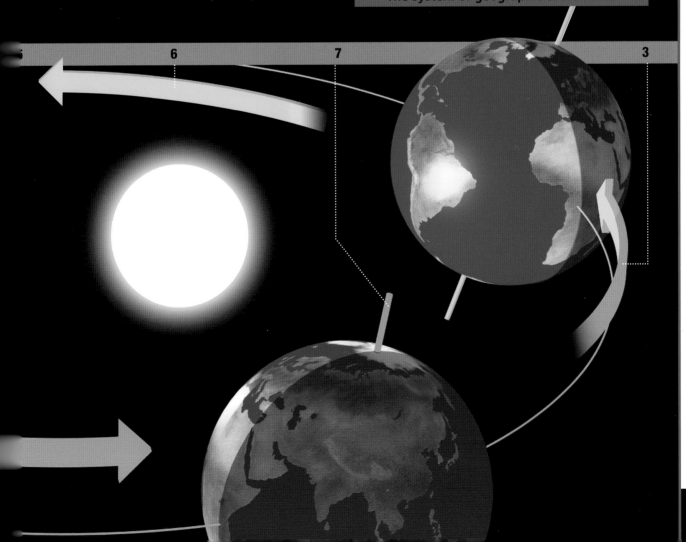

You can find the position of any point on Earth by drawing a series of imaginary rings parallel to the equator (parallels) and a series of perpendicular circles that converge at the poles (meridians).

summer solstice

opens from June 21 to June 22. The North Pole is at its closest point to the sun, so summer begins in the Northern Hemisphere, and winter begins in the south

spring equinox

occurs from March 20 to March 21. The solar rays fall perpendicularly on the equator, so that the lit-up surface is equal in both hemispheres. Day and night are of equal length. Spring begins in the Northern Hemisphere, and autumn begins in the south

6 fall equinox

takes place from September 22 to September 23. The sun's rays fall perpendicularly on the equator, so that the lit-up surface is equal in both hemispheres. Day and night are of the same length. Autumn begins in the Northern Hemisphere, and spring begins in the south.

7 axis of rotation

is the imaginary axis around which Earth spins itself; this axis occupies four positions during the course of the Earth's orbit around the sun. These positions define the four seasons of the year: spring, summer, autumn, and winter

1 longitude

is the angle measured along the equator of the Earth, between a reference meridian (the Greenwich meridian) and the meridian of the place

2 latitude

angle determined along the length of the meridian between a terrestrial point and the reference parallel (the equator)

3 polar axis
4 Arctic Circle
5 Antarctic Circle
6 Tropic of Cancer
7 Tropic of Capricorn
8 equator
9 Greenwich meridian
10 Northern Hemisphere
11 Southern Hemisphere
12 Eastern meridian
13 Western meridian

The system of geographic coordinates

A JOURNEY TO THE CENTER OF THE EARTH

Studies on the movement and velocity of seismic waves, combined with observations of surface rocks, volcanic lava, drilling, laboratory experiments, and meteorites, have proven that the Earth is structurally divided into different layers.

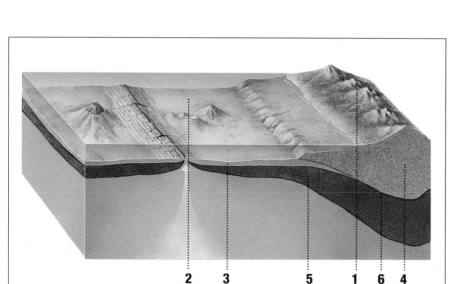

2   3   5   1   6   4

1 continental crust
between 12 and 31 miles (19 and 50 km) thick; consists of a sedimentary layer, a granite layer, and a basalt layer

2 oceanic crust
characterized by the absence of a granite layer; has a maximum thickness of 4 miles (6.4 km)

3 sedimentary layer
found on the continents and continental shelves

4 granite layer
makes up the basic mass of the continental areas

5 basalt layer
found in both continental and ocean areas

6 Conrad discontinuity
separates the granite and basalt layers of the crust

Structure of the crust

The crust is not a uniform layer. Its makeup and thickness vary according to whether it is oceanic crust or continental crust.

crust ■
is the outermost and thinnest layer

lithosphere ■
includes all of the crust and the top part of the mantle, forming plates that "float" over the asthenosphere

asthenosphere ■
semi-fluid layer located in the upper mantle, on top of which the lithosphere "floats"

outer core ■
made of nickel and iron in a molten state

inner core ■
made of nickel and iron; it is solid

Weichert-Lehman discontinuity ■
divides the outer core from the inner core

Gutenberg discontinuity ■
separates the lower mantle from the outer core

Repetti discontinuity ■
divides the upper mantle from the lower mantle

Mohorovicic discontinuity ■
separates the crust from the mantle

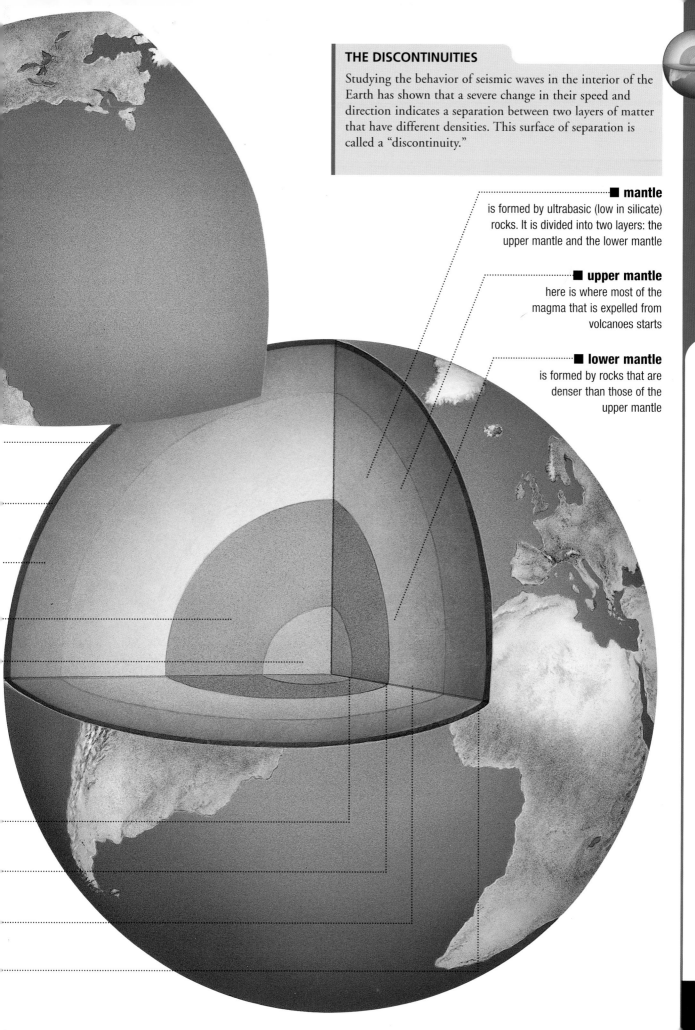

THE DISCONTINUITIES

Studying the behavior of seismic waves in the interior of the Earth has shown that a severe change in their speed and direction indicates a separation between two layers of matter that have different densities. This surface of separation is called a "discontinuity."

■ **mantle**
is formed by ultrabasic (low in silicate) rocks. It is divided into two layers: the upper mantle and the lower mantle

■ **upper mantle**
here is where most of the magma that is expelled from volcanoes starts

■ **lower mantle**
is formed by rocks that are denser than those of the upper mantle

THE AIR WE BREATHE

The atmosphere is a layer of gas that surrounds the Earth. It is formed mainly of nitrogen (78%), oxygen (21%), and much smaller quantities of other gases, such as CO_2 and water vapor. Air is essential for life, since it allows us to breathe, protects us by blocking dangerous solar radiation (ultraviolet rays), and keeps the surface of the planet from getting too hot or too cold. The atmosphere surrounds the Earth by the force of gravity. It is made up of a series of different layers.

THE OZONE LAYER

It is located in the stratosphere, approximately 9 to 31 miles (14.5 to 50 km) above the Earth's surface. Ozone is an unstable molecule of three oxygen atoms (O_3), which acts as a powerful solar filter that prevents a small part of ultraviolet radiation from reaching the Earth and harming living things.

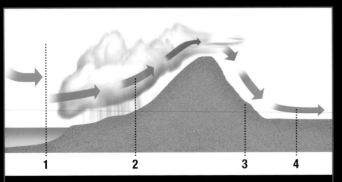

1 hot moist air

2 windward slopes, moist

3 leeward slopes, dry

4 dry wind (chinook)

Cloud formation

When the air ascends, it is cooled, and the water vapor it contains condenses to form little water droplets; this is what we call clouds. There are various ways for clouds to form.

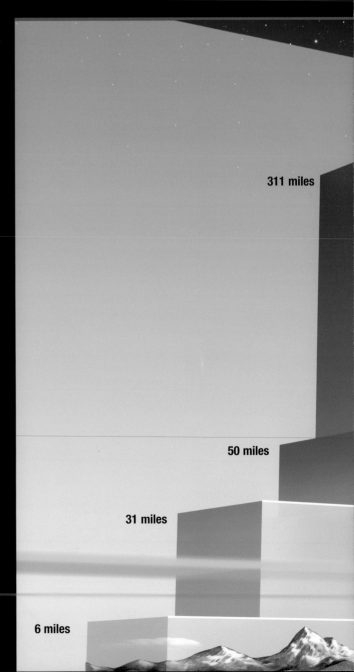

311 miles

50 miles

31 miles

6 miles

1 troposphere
is the densest layer, where weather phenomena take place; its composition allows life to develop

2 tropopause
border between the troposphere and the stratosphere

3 stratosphere
here is where the absorption of ultraviolet radiation takes place, thanks to the existence of the ozone layer

4 stratopause
is the border between the stratosphere and the mesosphere

5 mesosphere
the density of the air is very low, although the proportion of components is the same as in the troposphere; meteorites are destroyed when they smash against this layer

6 mesopause
border between the mesosphere and the thermosphere

7 thermosphere
the air density is very low, and the few atoms it contains are ionized; here the aurora borealis is produced, and this is where artificial satellites orbit

8 ionosphere
is not a separate layer, but rather makes up part of the thermosphere.

Long-distance radio communication is possible because the different parts of the ionosphere reflect radio waves back to Earth

9 thermopause
border between the thermosphere and the exosphere

10 exosphere
the lightest gases escape the pull of the Earth and are scattered in space; they may be detected up to 5,000 miles (8,047 km) from Earth

11 temperature
this band indicates the approximate temperature range of this area of the atmosphere

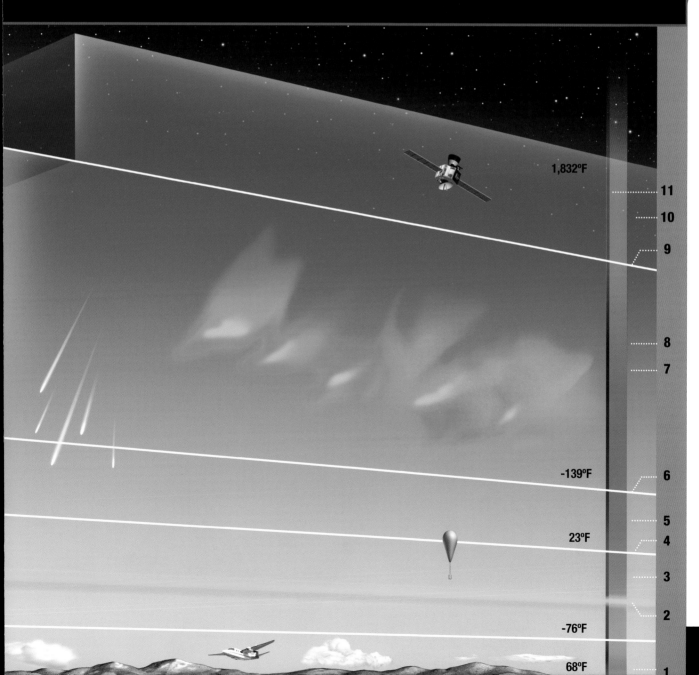

1,832°F

-139°F

23°F

-76°F

68°F

11
10
9

8
7

6

5
4

3

2

1

CONTINENTS ADRIFT

The crust and the upper part of the Earth's mantle make up the lithosphere. This is not a continuous layer. It is broken up into large pieces called tectonic plates, which move beside each other, "floating" on top of the semi-fluid layer underneath, the asthenosphere. The edges of the plates are unstable strips of great tectonic, seismic, and volcanic activity. This is where the Earth's crust is formed or destroyed.

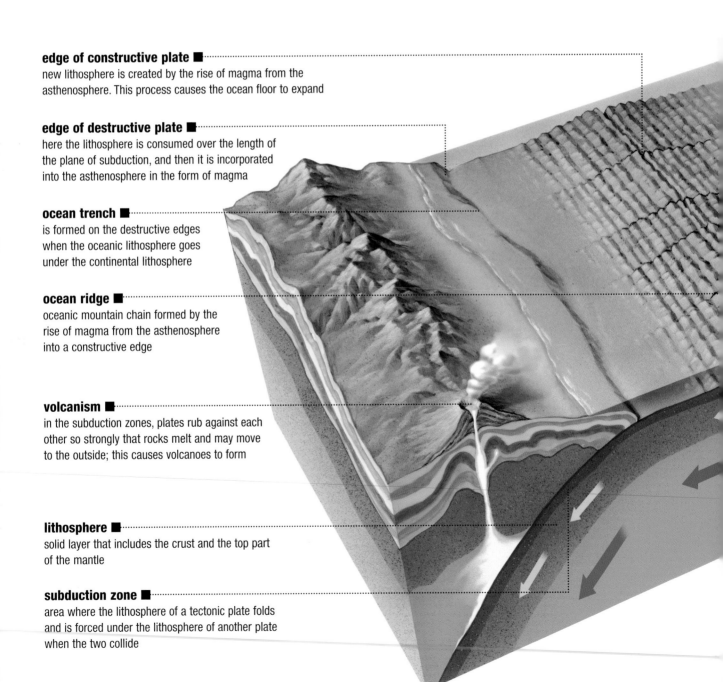

edge of constructive plate ■
new lithosphere is created by the rise of magma from the asthenosphere. This process causes the ocean floor to expand

edge of destructive plate ■
here the lithosphere is consumed over the length of the plane of subduction, and then it is incorporated into the asthenosphere in the form of magma

ocean trench ■
is formed on the destructive edges when the oceanic lithosphere goes under the continental lithosphere

ocean ridge ■
oceanic mountain chain formed by the rise of magma from the asthenosphere into a constructive edge

volcanism ■
in the subduction zones, plates rub against each other so strongly that rocks melt and may move to the outside; this causes volcanoes to form

lithosphere ■
solid layer that includes the crust and the top part of the mantle

subduction zone ■
area where the lithosphere of a tectonic plate folds and is forced under the lithosphere of another plate when the two collide

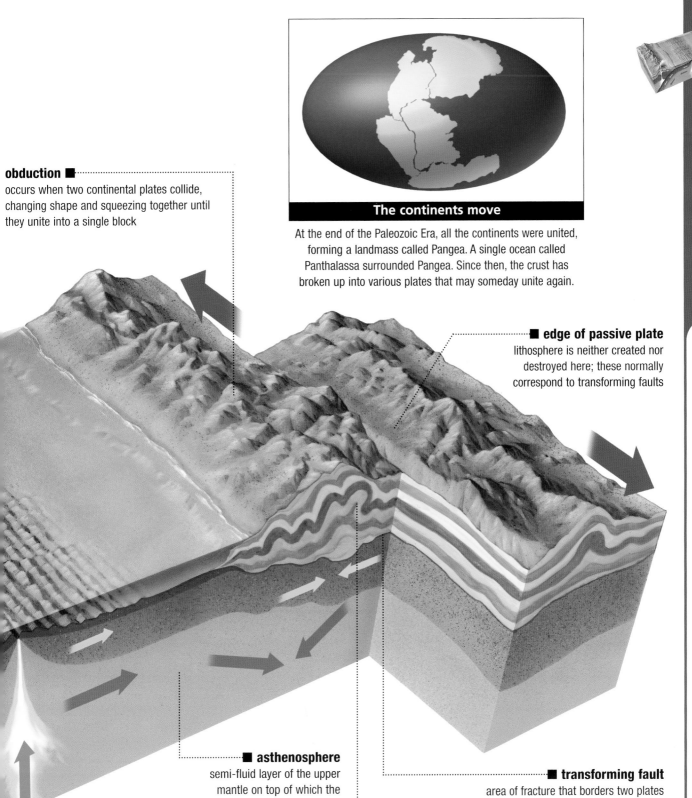

The continents move

At the end of the Paleozoic Era, all the continents were united, forming a landmass called Pangea. A single ocean called Panthalassa surrounded Pangea. Since then, the crust has broken up into various plates that may someday unite again.

obduction ■
occurs when two continental plates collide, changing shape and squeezing together until they unite into a single block

■ edge of passive plate
lithosphere is neither created nor destroyed here; these normally correspond to transforming faults

■ asthenosphere
semi-fluid layer of the upper mantle on top of which the lithosphere "floats"

■ transforming fault
area of fracture that borders two plates that are sliding sideways with respect to each other; this movement often causes earthquakes

THE OCEAN DEPTHS EXPAND

At the edges of constructive plates, there is a flow of materials coming from the upper mantle to the terrestrial (land) crust. In this manner, intrusive materials create new oceanic lithosphere and make the ocean bottom expand on both sides of the spines. This causes continental drift.

■ folding
when two plates of the continental lithosphere collide, both masses change shape and are squeezed together, which causes great mountain ranges to form

WHEN THE EARTH "SPITS" FIRE

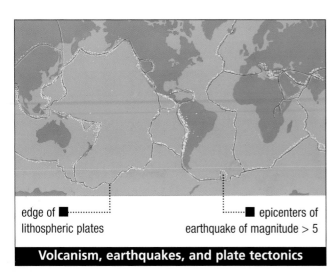

edge of ■ .... lithospheric plates

:........ ■ epicenters of earthquake of magnitude > 5

Volcanism, earthquakes, and plate tectonics

Most earthquakes and volcanoes are located at the edges of lithospheric plates. One of the most active areas is the "Pacific Ring of Fire."

Volcanoes form when molten material and gases from the interior of the Earth move up to the surface through fissures (cracks). This may happen in a violent manner or as a slow flow of lava. Over the last 10,000 years, 1,415 volcanoes have been active around the world. Some of them, such as those in Hawaii and Indonesia, and Mount Etna and Mount Stromboli in Italy, erupt frequently, while others remain dormant (inactive) for many years.

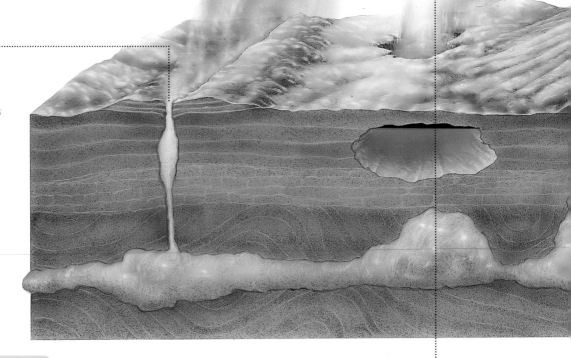

volcanic fissure ■
at times, lava takes advantage of fissures in the Earth's crust and flows in great quantities through these cracks

■ **geyser**
occasional eruptions of boiling water; these are actually water volcanoes that are produced when underground water comes into contact with magma

A KILLER VOLCANO

In August 1883, the volcano Krakatoa near the island of Java erupted, throwing up rocks to a height of 34 miles (55 km). The island on which it was located blew up with a force of 100 megatons (the atomic bomb that hit Hiroshima was about 20 kilotons). The blast was heard in Australia, and generated a tsunami 130 feet (40 meters) high. As a result of this natural disaster, 36,000 people died.

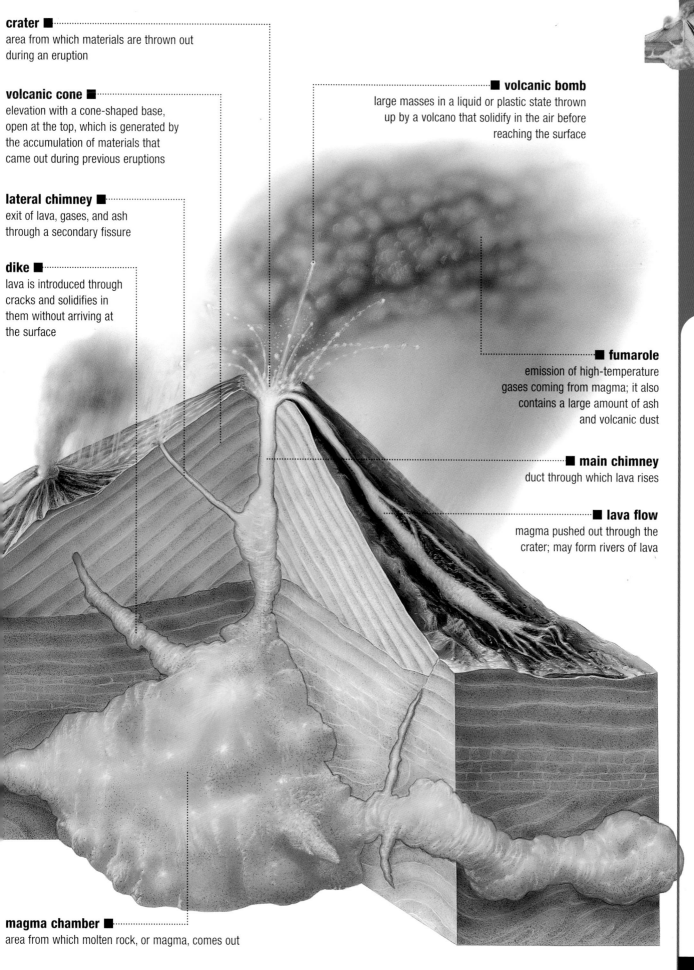

crater ■
area from which materials are thrown out
during an eruption

volcanic cone ■
elevation with a cone-shaped base,
open at the top, which is generated by
the accumulation of materials that
came out during previous eruptions

lateral chimney ■
exit of lava, gases, and ash
through a secondary fissure

dike ■
lava is introduced through
cracks and solidifies in
them without arriving at
the surface

■ **volcanic bomb**
large masses in a liquid or plastic state thrown
up by a volcano that solidify in the air before
reaching the surface

■ **fumarole**
emission of high-temperature
gases coming from magma; it also
contains a large amount of ash
and volcanic dust

■ **main chimney**
duct through which lava rises

■ **lava flow**
magma pushed out through the
crater; may form rivers of lava

magma chamber ■
area from which molten rock, or magma, comes out

THE WORLD UNDERGROUND

Karst formations are the result of limestone being dissolved by the action of rainwater and carbon dioxide. Although it is a compact rock, limestone has many irregularities in its surface—vertical fractures and stratification planes, for example. Runoff water gets into these irregular surfaces and dissolves the limestone. As more water seeps into the limestone, underground caves begin to appear.

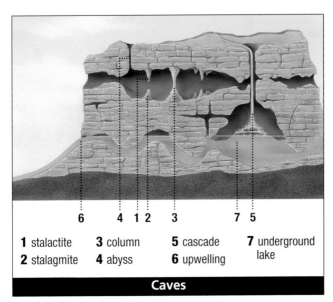

6　　4 1 2　3　　7 5

1 stalactite　**3** column　**5** cascade　**7** underground lake
2 stalagmite　**4** abyss　**6** upwelling

Caves

Caves are hollowed-out areas of rock that can be found both above and below the Earth's surface. As blocks fall from the ceiling of the hollowed-out area, the cave gradually widens. Inside, limestone not only dissolves, but is precipitated, forming strange and beautiful shapes.

upwelling ■
place where underground water appears on the surface

THE BIGGEST CAVE IN THE WORLD

The biggest cave in the world is in Sarawak (Malaysia). It has an area of 40 acres (16 hectares), a length of 2,300 feet (701 meters), an average width of 985 feet (300 meters), and its lowest part is 230 feet (70 meters) high.

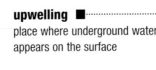

underground river ■
water takes advantage of the galleries and moves through them like an underground river

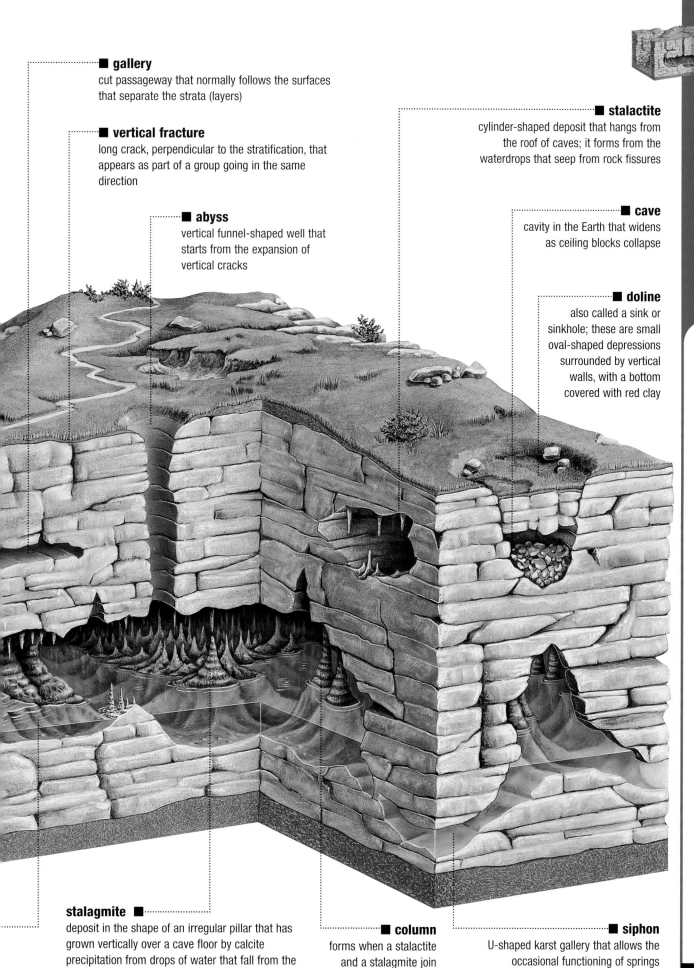

■ gallery
cut passageway that normally follows the surfaces that separate the strata (layers)

■ vertical fracture
long crack, perpendicular to the stratification, that appears as part of a group going in the same direction

■ abyss
vertical funnel-shaped well that starts from the expansion of vertical cracks

■ stalactite
cylinder-shaped deposit that hangs from the roof of caves; it forms from the waterdrops that seep from rock fissures

■ cave
cavity in the Earth that widens as ceiling blocks collapse

■ doline
also called a sink or sinkhole; these are small oval-shaped depressions surrounded by vertical walls, with a bottom covered with red clay

stalagmite ■
deposit in the shape of an irregular pillar that has grown vertically over a cave floor by calcite precipitation from drops of water that fall from the roof or from a stalactite

■ column
forms when a stalactite and a stalagmite join

■ siphon
U-shaped karst gallery that allows the occasional functioning of springs

RIVERS OF ICE

Glaciers are large masses of ice in motion that form in polar regions and in high mountains above the edge of perpetual clouds. They are impressive phenomena, able to reduce rock to dust. They carry millions of tons of sediments across great distances. To form, glaciers must accumulate a large amount of snow, which is compacted and has enough slope to make it move on one side.

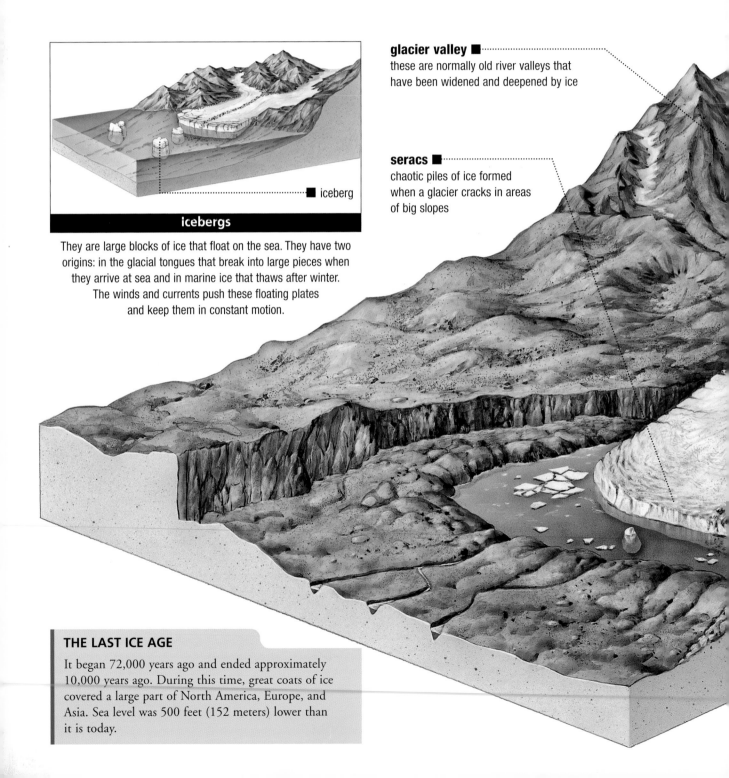

glacier valley ■
these are normally old river valleys that have been widened and deepened by ice

seracs ■
chaotic piles of ice formed when a glacier cracks in areas of big slopes

■ iceberg

icebergs

They are large blocks of ice that float on the sea. They have two origins: in the glacial tongues that break into large pieces when they arrive at sea and in marine ice that thaws after winter. The winds and currents push these floating plates and keep them in constant motion.

THE LAST ICE AGE

It began 72,000 years ago and ended approximately 10,000 years ago. During this time, great coats of ice covered a large part of North America, Europe, and Asia. Sea level was 500 feet (152 meters) lower than it is today.

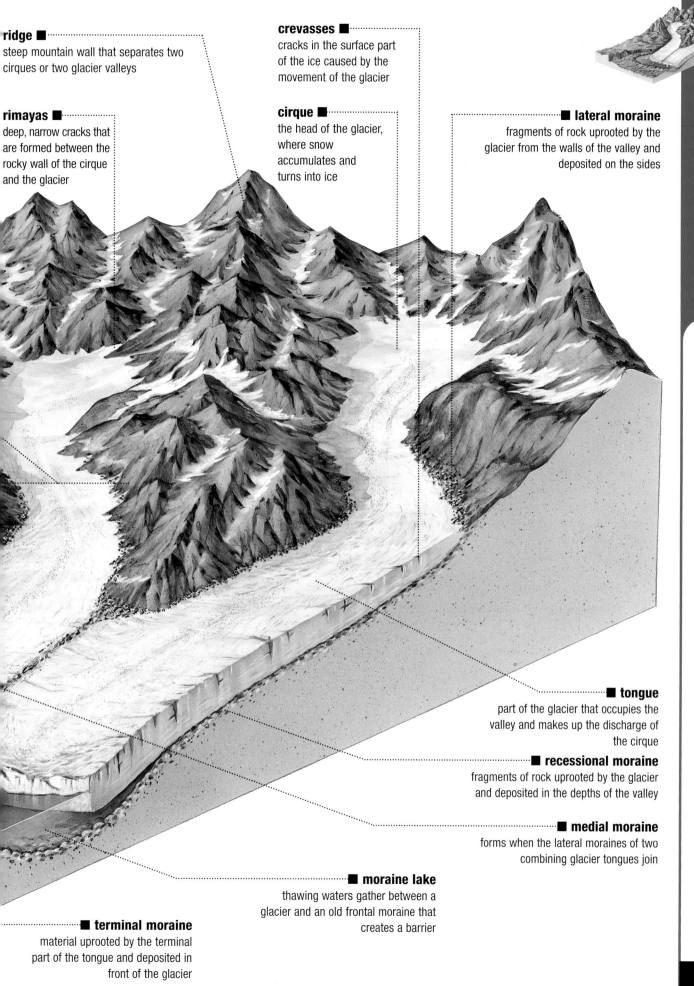

ridge ■
steep mountain wall that separates two cirques or two glacier valleys

rimayas ■
deep, narrow cracks that are formed between the rocky wall of the cirque and the glacier

crevasses ■
cracks in the surface part of the ice caused by the movement of the glacier

cirque ■
the head of the glacier, where snow accumulates and turns into ice

■ **lateral moraine**
fragments of rock uprooted by the glacier from the walls of the valley and deposited on the sides

■ **tongue**
part of the glacier that occupies the valley and makes up the discharge of the cirque

■ **recessional moraine**
fragments of rock uprooted by the glacier and deposited in the depths of the valley

■ **medial moraine**
forms when the lateral moraines of two combining glacier tongues join

■ **moraine lake**
thawing waters gather between a glacier and an old frontal moraine that creates a barrier

■ **terminal moraine**
material uprooted by the terminal part of the tongue and deposited in front of the glacier

A WORLD UNDER WATER

Most of the Earth is covered by water, almost all of it belonging to the seas and oceans. Some of these bodies of water are largely unexplored and as hostile as outer space, due to the sizes and depths of certain areas. The bottom of the sea varies as much as dry land. There, we find mountain ranges, volcanoes, precipices, and other features.

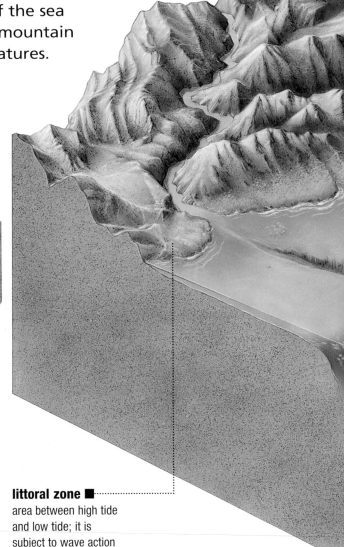

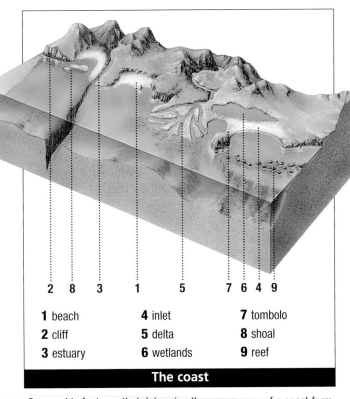

1 beach	**4** inlet	**7** tombolo
2 cliff	**5** delta	**8** shoal
3 estuary	**6** wetlands	**9** reef

The coast

Geographic features that determine the appearance of a coast form because of the action of waves, tides, and currents.

littoral zone ■
area between high tide and low tide; it is subject to wave action

undersea canyon ■
old submerged river valley reexcavated by the currents of mud produced here

CURRENTS

These are periodic rises and descents of the seawater level that happen every 12 hours and 26 minutes, due mainly to the pull of the moon and the sun on the liquid mass of the sea. The maximum level is known as high tide, and the minimum level is called low tide.

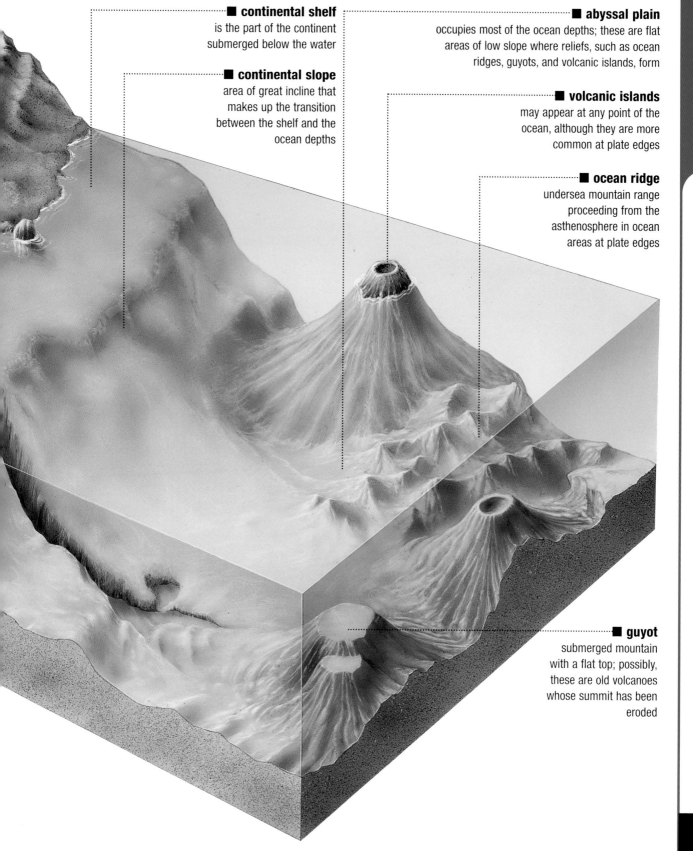

neritic zone
is the area included between the low tide and the continental slope

pelagic zone
extends from the edge of the continental shelf and covers the deepest areas

■ **continental shelf**
is the part of the continent submerged below the water

■ **continental slope**
area of great incline that makes up the transition between the shelf and the ocean depths

■ **abyssal plain**
occupies most of the ocean depths; these are flat areas of low slope where reliefs, such as ocean ridges, guyots, and volcanic islands, form

■ **volcanic islands**
may appear at any point of the ocean, although they are more common at plate edges

■ **ocean ridge**
undersea mountain range proceeding from the asthenosphere in ocean areas at plate edges

■ **guyot**
submerged mountain with a flat top; possibly, these are old volcanoes whose summit has been eroded

OUR EVER-CHANGING PLANET

The Earth is in a continuous process of transformation because of the combined action of two forces that act on the terrestrial crust. The internal force is constructive and the external force is destructive. These same processes have been active for hundreds of millions of years. Internal forces are manifested through volcanoes, earthquakes, folding, and sinking. The most obvious external force is erosion.

1 horizon A
rich in organic matter; vegetation is rooted here

2 horizon B
almost without organic matter; here substances dragged from above are precipitated

3 horizon C
formed by meteorized fragments of bedrock

4 horizon D
is unaltered rock; marks the lower limit of the soil

Soil

This is the uppermost layer of the continental crust. It is formed by rocks, air, water, organic matter, and living things.

THE ACTION OF LIVING THINGS

Plant roots help widen cracks in rocks and break them apart, making it easier for water and air to penetrate them. But human beings are most to blame for the rapidly changing environment. We are consistently altering nature by building roads and dams, digging mines, and cutting down forests.

sedimentation ■
the transported materials are deposited in the most depressed areas of the terrestrial crust

meteorization ■
the materials are changed by the chemical effects of water and air

subsidence ■
constant sedimentation produces an increased load, which leads to the burial of sediments

metamorphism ■
increases in pressure and temperature cause mineralogical changes in the rock, which create metamorphic rock

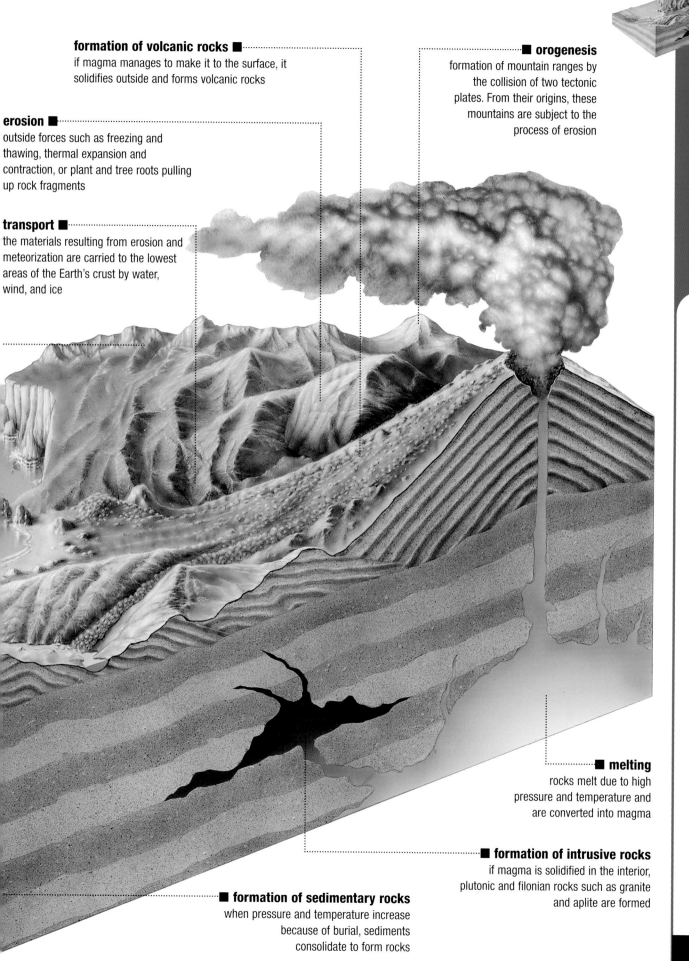

formation of volcanic rocks ■
if magma manages to make it to the surface, it solidifies outside and forms volcanic rocks

erosion ■
outside forces such as freezing and thawing, thermal expansion and contraction, or plant and tree roots pulling up rock fragments

transport ■
the materials resulting from erosion and meteorization are carried to the lowest areas of the Earth's crust by water, wind, and ice

■ orogenesis
formation of mountain ranges by the collision of two tectonic plates. From their origins, these mountains are subject to the process of erosion

■ melting
rocks melt due to high pressure and temperature and are converted into magma

■ formation of intrusive rocks
if magma is solidified in the interior, plutonic and filonian rocks such as granite and aplite are formed

■ formation of sedimentary rocks
when pressure and temperature increase because of burial, sediments consolidate to form rocks

MAP MAKING

A map is a two-dimensional (length and width) representation of a region, but the Earth and its topographic features have three dimensions. The third dimension (altitude) is represented by curves that show level. These curves are lines that connect relief points located at equal height over sea level. To represent a region on a map, it is assumed that the relief is intersected by a series of equally distant planes, whose projection yields the tracing of level curves.

4 escarpment ■

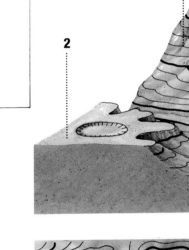

3 steep slope ■

2 plateau ■

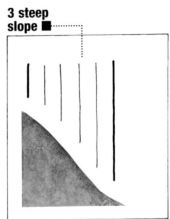

1 valley ■

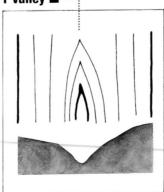

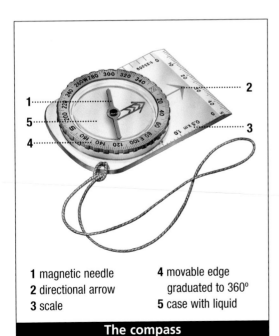

1 magnetic needle
2 directional arrow
3 scale
4 movable edge graduated to 360°
5 case with liquid

The compass

It is an instrument for determining direction. A compass has a needle that is oriented with the force lines of the Earth's magnetic field. It provides an indication of direction with respect to the Earth's magnetic north.

WHERE IS NORTH?

There are three types of north: geographic north (which is the intersection point between the Earth's axis of rotation and its surface), magnetic north (which is what a compass indicates), and the third north (which is shown on a map). The difference between the first two is known as magnetic declination, and its value depends on the time zone of a particular location.

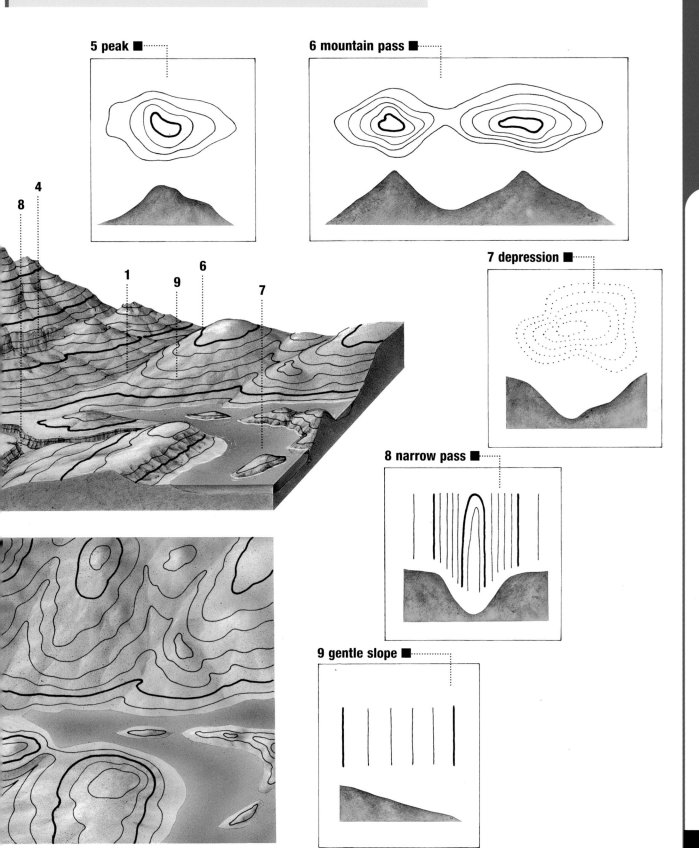

5 peak ■

6 mountain pass ■

7 depression ■

8 narrow pass ■

9 gentle slope ■

4

8

1

9

6

7

DID YOU KNOW?

Era	Period		Characteristic Fauna		MA*
PHANEROZOIC					
Cenozoic	Quaternary		Present-day fauna	Era of humankind	
			Primitive hominids		
			Mammoths		1.8
	Tertiary	Neogenic	Mammals	Era of mammals	
			Lamellibranchs		
			Freshwater shellfish		23.8
		Paleogenic	Nummulites		
			Mammals		
			Echinoderms (sea urchins)		65
Mesozoic	Cretaceous		Ammonites	Era of reptiles and ammonites	
			Foraminifers		
			Dinosaurs		
			Pachyodonts		144
	Jurassic		Ammonites		
			Belemnites		
			Dinosaurs		
			Birds with teeth		206
	Triassic		Ammonites		
			Lamellibranchs		
			Crinoids		
			Reptiles (dinosaurs)		248
Paleozoic	Permian		Amphibians	Era of trilobites	
			First reptiles		
			Brachiopods		290
	Carboniferous		Goniatites		
			Brachiopods		
			Fusulinas		
			Amphibians		360
	Devonian		Armored fish		
			Ostracods		
			Goniatites		
			Corals		
			Brachiopods		409

Era	Period	Characteristic Fauna		MA*
PHANEROZOIC Paleozoic	Silurian	Graptolites	Era of trilobites	
		Conodonts		439
	Ordovician	Graptolites		
		Brachiopods		
		Gasteropods		510
	Cambrian	Trilobites		
		Brachiopods		
		Archeocytes (calcareous sponges)		570
CRYPTOZOIC Precambrian	Proterozoic	Invertebrates without a hard skeleton		
		First multicellular algae		
		First protozoa		
		First multicellular organisms		2,500
	Archaic	First eukaryotic cells (bacteria)		3,900
	Prearchaic	Without life		4,600

* Millions of Years Ago

GEOLOGICAL TIME

Geological processes generally occur so slowly that humans cannot see them happen. For this reason, geologists over the last century have developed a time scale based on global geological and biological events that are used as sequential reference points. This scale, counting backward in millions of years, starts with the time of the Earth's formation. The periods are established on the basis of geological and biological criteria. The large periods have a planetary reach and are the bases for establishing geological time.

THE EARTH IN FIGURES

Mass (lbs)	1.32×10^{25}
Equatorial radius (miles)	3,963.19
Average density (lbs/cm³)	12.19
Average distance to the sun (miles)	92,957,130.4
Rotational period (days)	0.99727
Rotational period (hours)	23.9345
Orbital period (days)	365.256
Average orbital velocity (miles/sec)	18.5
Inclination of axis	23.450°
Surface gravity at the equator (feet/sec²)	32
Temperature	⁻128°F–136°F
Average surface temperature	59°F
Atmospheric pressure (bars)	1.013
Atmospheric composition	
Nitrogen	77%
Oxygen	21%
Others	2%

EARTHQUAKES

Earthquakes are produced when the tension accumulated by the deformation of the Earth's layers is suddenly released. Rock masses subjected to gigantic forces are broken, reorganizing matter and releasing huge amounts of energy, which causes the Earth to tremble. The initial foci (epicenters) of earthquakes are found at different depths: The deepest may reach up to 435 miles (700 km). Earthquakes are especially frequent near the edges of tectonic plates.

These jolts are very hard to predict. Today, there are no effective systems to warn people ahead of time about a coming seismic event.

COMPARISON TABLE BETWEEN THE MERCALLI SCALE AND THE RICHTER SCALE

Mercalli Scale (intensity)	Richter Scale (magnitude)	Observations
I	Up to 2.5 instrumental	Weak seismic event registered only by seismographs.
II	From 2.5 to 3.1 very weak	Perceived only by persons at rest.
III	From 3.1 to 3.7 light	Perceived in densely populated areas by part of the population.
IV	From 3.7 to 4.3 moderate	Felt by people in movement; some sleeping people awaken.
V	From 4.3 to 4.9 somewhat strong	Felt outside; people wake up.
VI	From 4.9 to 5.5 strong	Perceived by everyone; walking unstable; trees and materials are shaken by the effect of the seismic event.
VII	From 5.5 to 6.1 very strong	Difficulty staying on one's feet; hanging objects fall; there may be small collapses and slides.
VIII	From 6.1 to 6.7 destructive	Partial collapse of structures; considerable damage in ordinary buildings.
IX	From 6.7 to 7.3 ruinous	Considerable damage in specially constructed buildings; complete collapse of edifices and houses; general damage in cement structures, dams, and dikes.
X	From 7.3 to 7.9 disastrous	Destruction of most buildings; bridge collapses; serious damage in dams and piers.
XI	From 7.9 to 8.4 very disastrous	Few structures remain standing; large fissures in the terrain.
XII	From 8.4 to 9 catastrophic	Total destruction; great masses of rock displaced; objects thrown in the air.

INDEX